Blood of Mars

A Story by

Andrew Polykandriotis

"Loyalty is a choice. Betrayal is forever."

Dedication

"To my family, for their unwavering support through my obsessions, my creativity, and my madness. For every late night I spent lost in worlds of my own making, thank you for believing in me."

Summary

"Blood of Mars" is a gripping sci-fi epic of power, betrayal, and war, where Princess Ankhes Solis must choose between loyalty to her father's crumbling empire or the exiled brother who seeks to reshape Mars. As rival factions prepare for battle and Earth tightens its grip, an ancient force stirs beneath the planet's surface, threatening them all.

Teshan nekhet!   (Victory in blood)

Table of Contents

Blood of Mars - Chapter 1: Prologue

The twin moons of Phobos and Deimos projected weak, ghostly shadows across the Martian sky, which flared in an ethereal blend of deep orange and scarlet. Nature and technology came together in the northern polar area to create an unmatched spectacle—a feat made possible by blood, sacrifice, and unwavering ambition. Beyond the horizon, towering emerald-green trees extended, their roots fed by rivers that cut through the rocky terrain like life's veins. Under the starlight, the transparent structures of enormous domed cities pulsated with artificial life, resembling shimmering monoliths against the wide Martian plains. This appeared to be heaven to the inexperienced eye.

But to Emperor Ramses Solis II, it was not enough.

From the observation platform of the Citadel of Nefer'Solis, he surveyed his empire, standing tall and unyielding, his silhouette framed against the endless vista of conquest. The chamber around him was an architectural marvel—a seamless blend of ancient Martian hieroglyphics and the most advanced technology ever conceived. Golden columns, etched with the storied rise of the Solis dynasty, lined the walls. The air hummed faintly with the soft murmur of life support systems, and the scent of polished metal and exotic incense filled the space.

Ramses' emerald-green eyes, piercing and analytical, reflected the flickering city lights below. His expression was inscrutable—cold, calculating, a ruler devoid of weakness. At six foot ten, his presence alone commanded absolute obedience. Draped in robes of crimson and gold, the colors shimmered like molten fire, a testament to his divine right to rule. He was not just an emperor; he was a god among men, the product of ten generations of Martian supremacy.

Behind him, the massive iron doors of the throne room groaned open, breaking the serene silence.

Ramses did not turn. He already knew who had entered.

"Your Eminence."

The voice belonged to Setekh Solis, the eldest of Ramses' eleven sons. Sharp, deliberate, and tinged with the cold precision of a seasoned warrior. Setekh had spilled countless gallons of blood in service to the empire. He had burned cities, crushed rebellions, and carved his name into the annals of history—but here, before his father, he was merely a son.

"The southern colonies stir once more," Setekh reported. "Coordinated raids have struck the Sekhet District. Supplies stolen. Two Martian sentries executed."

Ramses clenched his jaw.

"The South," he muttered. Then, after a pause, he spoke the name as if it were venom on his tongue. "Kael."

A silence stretched between them. The weight of the name hung in the air like a curse.

Only now did Ramses turn, his emerald eyes locking onto his son. Setekh stood firm, his golden armor polished to a faultless sheen, but even he—battle-hardened, bloodied by war—felt the oppressive gravity of his father's gaze.

"How many times must I crush him beneath my heel?" Ramses' voice was like the slow, deliberate grinding of tectonic plates. "Kael forgets his place. He forgets the blood in his veins."

Setekh hesitated. His fingers twitched at his side, betraying the sliver of unease he dared not voice.

"Kael is… different, Father," Setekh admitted. "His ties to the Earthlings have given him a following. To them, he is a savior."

A savior.

Ramses exhaled sharply, his patience wearing thin. He stepped forward, closing the space between them, his presence an unrelenting storm pressing against Setekh's very being.

"Kael is an abomination," he snarled. "A half-breed mistake. A stain upon the bloodline of Solis."

Setekh said nothing. He had heard these words before, but never had his father spoken them with such barely contained fury.

"What would you have me do, Father?" he asked cautiously. "Thutmose suggests a preemptive strike—starve them before they can rally."

Ramses turned away, his hands clasping behind his back as he gazed once more upon his empire.

"Thutmose is a hammer. He sees every problem as a nail to be crushed."

His voice dropped to a whisper, though the venom remained.

"No. Lucius will take the lead. Send him south."

Setekh blinked, momentarily taken aback. Lucius. Of all his brothers, Lucius was the most ruthless. Where Setekh wielded strategy, and Thutmose brute force, Lucius was the blade in the dark, the whisper before the dagger found its mark.

"As you command, Father."

Setekh turned to leave, but Ramses spoke once more.

"And Setekh."

The eldest son halted.

"Ensure that Ankhes is present for the council meeting tomorrow. She must learn the cost of betrayal."

A pause.

Setekh exhaled slowly, his expression unreadable.

"Yes, Father."

He bowed once and exited the chamber, leaving Ramses alone with his thoughts.

The emperor's gaze flickered toward the southern horizon.

Kael... you will learn what it means to defy me.

The Princess of Mars

Deep within the royal palace, far removed from the weight of war and politics, Ankhes Solis sat alone in her private chamber.

Her quarters were a masterpiece of design, adorned in gold and crimson, with intricate hieroglyphic carvings chronicling the Solis dynasty's dominance over Mars. A large, circular window overlooked the sprawling northern forests, an ocean of emerald green stretching endlessly toward the horizon.

She stood at the window's edge, fingers tracing the cool glass. Her amber eyes, luminous even in the dim lighting, flickered with restlessness. Her dark curls cascaded over her shoulders, framing her sharp, high-cheekboned face—the very image of Martian nobility.

But she was troubled.

Whispers of rebellion traveled through the palace like the echoes of a coming storm. Her father's wrath was an inevitability, and Kael—the brother she had not seen in years, the brother cast out like a diseased limb—was at its epicenter.

A knock at the door shattered her thoughts.

"Enter," she called.

The door swung open to reveal Lucius Solis.

Tall, regal, and sharpened like a blade, Lucius bore the hallmark of a man who thrived in shadows and secrets. His golden armor caught the light as he smirked, though the gesture was devoid of warmth.

“Sister,” he greeted smoothly. “Father has summoned you for tomorrow’s council meeting.”

Ankhes arched an eyebrow. “What matter is so urgent?”

“Kael.”

Lucius’ smirk widened as he watched her reaction.

“Father believes it’s time for you to understand the price of disloyalty.”

Ankhes turned back to the window, her expression unreadable.

“Disloyalty,” she murmured. “An interesting word, coming from him.”

Lucius stiffened slightly but said nothing.

“Do not let Father hear you speak such things,” he warned. “His patience is not infinite.”

She turned, locking eyes with him.

“Thank you for your concern, brother,” she said coolly. “You may leave now.”

Lucius hesitated but inclined his head before stepping away. The door clicked shut, leaving Ankhes in silence.

She exhaled slowly.

Tomorrow would bring the council meeting.

And with it, the beginning of a storm that would change the fate of Mars forever.

Blood of Mars - Chapter 2: Princess Ankhes Solis

The Enigmatic Heir

The Martian capital, Nefer'Solis, gleamed like a black jewel against the vast crimson desert. Obsidian spires clawed toward the heavens, their golden hieroglyphs chronicling centuries of conquest and dominion. At the city's heart stood the Citadel, an architectural marvel of gold and black stone, a fortress that housed the ruling dynasty of Mars.

In a chamber bathed in the eerie glow of Phobos and Deimos, Princess Ankhes Solis sat in silence, lost in thought.

She was the sole heir to the Martian Empire, the daughter of Emperor Ramses Solis II. To the world, she was the pinnacle of Martian perfection—tall, statuesque, with piercing amber eyes that burned like fire. Her crimson robes, embroidered with gold, framed her in an aura of divine authority.

Yet beneath her flawless exterior, she was restless.

She had been trained to rule, to uphold the purity of the Solis bloodline, to be unyielding. But Ankhes questioned the empire's rigid hierarchy. She questioned her father's obsession with bloodline purity, and most of all, she questioned the exile of her brother, Kael.

His name had become a whisper in the halls of the Citadel.

But now, his rebellion in the South was growing louder.

A knock at the door snapped her from her thoughts.

"Enter."

The heavy doors swung open, revealing Amon Rahar, the Chief of the Princess's Guard.

Amon was a formidable presence—tall, broad-shouldered, with the quiet confidence of a man who had killed in the name of duty.His black and gold armor was perfectly fitted, a silent reminder of his unwavering devotion.

But to Ankhes, Amon was more than just a bodyguard.

He was her confidant. Her lover. Her most trusted shadow.

"You summoned me, my princess?" Amon asked, bowing deeply.

Ankhes studied him for a long moment, amber eyes flickering with something unreadable.

"I did," she finally said. "I need your loyalty tonight more than ever."

Amon's gaze did not waver. "You have always had it."

A slow, knowing smile touched her lips.

"Good," she murmured. "Then prove it."

Testing Loyalty

She led Amon into her private chamber, a sanctuary few had ever entered.

The walls were carved with ancient Martian symbols, glowing faintly under the flickering red light of braziers. The air was thick with the scent of sandalwood and rare incense, an intoxicating blend of power and seduction.

She turned to face him.

"You know what I expect, Amon."

Amon knelt without hesitation, lowering his head. It was a ritual, an unspoken test of loyalty, a silent vow that bound them together.

Ankhes closed her eyes, allowing the moment to settle over her, allowing it to ground her in the reality of what she was about to do.

When it was over, she raised him to his feet, her voice calm, controlled.

"Amon, I have a mission for you."

"Anything, my princess."

"Kael's mother still lives in the North. Find her. Arrest her. Bring her to me—alive. And tell no one. Especially my father."

For the first time, Amon hesitated.

"My princess…" he said carefully. "If your father—"

"You swore loyalty to me," Ankhes interrupted, stepping closer. "Or was that a lie?"

Amon's jaw tightened.

"No, my princess. I will do as you command."

Ankhes nodded once. "Then go."

The Interrogation

The old woman was brought to Ankhes' private dungeon under the cover of darkness.

She was a ghost of the past, her once-beautiful face now lined with hardship. But despite her frail frame, her eyes burned with defiance as she was dragged into the interrogation chamber.

Ankhes stood before her, a towering contrast of youth and power.

"You know who I am," Ankhes said coldly.

Kael's mother lifted her chin.

"You are the Emperor's puppet," she spat. "A hollow shadow of the man you call father. But you are nothing compared to Kael."

A flicker of anger sparked in Ankhes' chest, but she pushed it down.

"You will tell me everything—his exile, his rebellion. You will tell me the truth."

Kael's mother laughed, bitter and sharp.

"The truth?" she said. "Your father wouldn't know truth if it struck him dead."

Ankhes stepped forward, voice like steel. "Speak. Now."

Kael's mother exhaled slowly. And then, she spoke.

She told of Kael's exile, of Ramses' obsession with purity, of betrayal disguised as justice.

But then—her voice shifted, something new seeping into her words.

A secret.

A truth Ankhes had never expected.

"I lied to your father," Kael's mother whispered. "Kael is not his son. His true father is Amon Rahar."

The Shocking Revelation

The room tilted.

Ankhes froze, her mind reeling.

She turned to Amon, who stood motionless in the shadows.

"Is it true?"

Amon's face was pale. He dropped to one knee.

"My princess," he whispered. "I beg your forgiveness. I—"

"Enough."

Ankhes' voice was cold now, detached. The man she had trusted above all others, the man who had shared her bed and her secrets, was a traitor.

She turned back to Kael's mother, her expression unreadable.

"You will pay for your lies."

Kael's mother laughed softly, shaking her head.

"Do what you will," she said. "But Kael will come for you. And when he does, this empire will burn."

Betrayal and Brutality

Ankhes exhaled slowly.

"Amon."

He lifted his head.

"Prove your loyalty to me."

Silence.

"Kill her."

Amon did not move.

For the first time, Ankhes saw hesitation in his usually unreadable expression.

"She is Kael's mother," he whispered.

"And you are my sword," Ankhes countered. "Do it."

A long pause.

Then—a blade flashed.

The execution was swift but brutal.

Kael's mother crumpled, her last breath escaping in a whisper.

"Kael will come for you."

The room fell silent.

Ankhes turned to Amon.

"Dispose of the body," she said, voice devoid of emotion. "And remember, Amon—your life depends on your loyalty to me."

The Aftermath

As the chamber emptied, Ankhes stood alone.

Her mind was a storm of realization.

She had crossed a line.

There was no turning back.

Blood of Mars - Chapter 3: The Banished Prince

The frozen wastelands of the Martian South Pole stretched infinitely, a harsh desert of ice and blood-red sands. The harsh winds howled relentlessly, whipping across the barren landscape. Yet, within this inhospitable domain, a thriving city stood in defiance of the Emperor's iron grip. It was a beacon for outcasts, Earth-born exiles, and Martian defectors alike—a place where the Empire's reach could not extend. And at its heart was Kael Solis, the banished son of Mars.

Kael, towering at nearly seven feet with broad shoulders and the unmistakable amber eyes of the Solis bloodline, was not just the leader of the South but its heart. He was a contradiction: ruthless but just, manipulative but compassionate, driven by vengeance but tempered by survival. His hybrid nature—part Martian, part Earthling—only added to his allure. Women, whether born under the Empire's red skies or Earth's blue oceans, could not resist him. His charisma and intelligence made him a dangerous adversary to his father, the Emperor, and a compelling savior to the downtrodden.

But Kael carried secrets darker than the frigid Martian nights.

The Banished Son

Kael had been banished from the Empire six years earlier. The memories of that day were burned into his mind as vividly as the twin Martian moons. His crime? Bedding the Emperor's youngest and most prized concubine—a Martian beauty no older than 19 summers. She had been the Emperor's newest acquisition, and her infatuation with Kael had been immediate. For weeks, they had met in secret, stealing moments in the labyrinthine halls of the Citadel.

Kael's charm was impossible to resist. His confidence, his wit, and the magnetic intensity of his amber gaze could make even the most loyal subjects betray their oaths. But it wasn't just the Emperor's concubine who had fallen for him. Kael had seduced the wife of one of his own brothers, further igniting the firestorm that would lead to his exile.

On the day of his banishment, the Emperor had stood before him, flanked by his council and guards. "You are a disgrace to the Solis name," Ramses had thundered, his voice echoing through the throne room. "Your lust, your arrogance—it poisons this empire. Were it not for your blood, I would strike you down where you stand."

Kael had met his father's gaze with defiance. "You fear me," he had said, his voice calm and unwavering. "You fear my knowledge, my power. That is why you won't kill me."

Ramses had been silent for a long moment before speaking again. "Get out of my sight," he had hissed. "If you ever set foot in the North again, I will kill you myself."

A New Kind of Leader

Exiled to the South, Kael had built something remarkable. His hybrid nature—the result of his mother's Earthling blood—made him a bridge between two worlds. While the Empire viewed Earthlings as inferior, Kael saw their potential. He welcomed defectors and rebels into his city, forging a new society that combined the ingenuity of Earth with the strength of Mars.

But Kael was no saint. He ruled with an iron fist, ensuring loyalty through both fear and respect. He knew that his father's spies were everywhere, waiting for the slightest sign of weakness. And so, he kept his plans close, sharing them only with his most trusted advisors.

Among those plans was his most ambitious yet: to weaponize the Empire's terraforming technology against them.

The Terraforming Bomb

Terraforming had transformed Mars from a barren wasteland into a world capable of supporting life. At the heart of this process was the Genesis Reactor, a technological marvel capable of igniting the chemical reactions necessary to create an atmosphere. But Kael, with his unparalleled understanding of Martian technology, had discovered something chilling: the Genesis Reactor could also be used as a weapon.

By destabilizing the device's core, it could release a wave of destructive radiation, obliterating everything in its path. Kael's engineers had already built several of these modified devices, which they called Crimson Flares, in secret. All that remained was acquiring the fuel to power them.

Kael knew the stakes. Deploying the Crimson Flares would mean war—total, unrelenting war. And yet, he hesitated. Despite his hatred for his father, there was one person who still tethered him to the Empire: his sister, Princess Ankhes.

A Message to the Princess

Kael stood in the war room of his fortress, surrounded by maps, schematics, and projections of the Empire's cities and defenses. His most trusted advisor, a grizzled Earthling named Mara, stood at his side.

"You're hesitating," Mara said, her voice low but firm.

Kael sighed, running a hand through his dark hair. "I'm not hesitating," he said. "I'm calculating. There's a difference."

"You want to warn her," Mara said. It wasn't a question.

Kael didn't respond immediately. Instead, he stared at the holographic map of the Empire, his gaze fixed on the northern polar region. "She deserves to know," he said finally. "Ankhes is… different. She's not like him."

Mara crossed her arms. "And what happens if she tells the Emperor?"

"She won't," Kael said. "I know her. She's loyal to him, yes, but she's also… curious. She'll want to hear me out."

Mara shook her head. "You're risking everything."

Kael turned to her, his amber eyes burning with intensity. "This isn't just about her. It's about the future of Mars. If she can be turned—if she can join us—then we have a chance to unite this planet."

Mara didn't argue further. She knew better than to try.

Kael summoned one of his spies, a Martian woman who had infiltrated the Empire years earlier. Her name was Selene, and she was as cunning as she was beautiful.

"You have a message for me, my lord?" Selene asked, bowing slightly.

"Yes," Kael said. "You're to return to the Citadel. Deliver this to my sister. Make sure she reads it in private." He handed her a sealed envelope, marked with the sigil of the South.

"And if she doesn't?" Selene asked.

"Then convince her," Kael said, his tone leaving no room for argument. "Do whatever it takes."

The Message

The message Selene delivered to Ankhes was simple but powerful. It contained a warning:

"Ankhes,

The Empire's destruction is imminent. The tools our father used to build his kingdom will be the same tools that tear it apart. But it doesn't have to end this way. Come to the South. See for yourself what we've built. Help me unite the poles—unite our people.

If you trust me, meet me at the southern citadel. Come alone.

Kael"

A Fateful Decision

As Kael watched Selene disappear into the icy night, he felt a flicker of hope. He had spent years preparing for war, but deep down, he wanted something more: peace, unity, and a chance to prove that Mars could be more than just a fractured planet ruled by fear.

But Kael was no fool. He knew that Ankhes' decision, whatever it might be, would set the course for Mars' future. And he was prepared for whatever came next.

For now, he waited.

Blood of Mars - Chapter 4: The Eleven Brothers

The Council Hall of the Citadel was an imposing chamber, a cavernous space filled with echoes of footsteps and voices that seemed to vanish into the towering obsidian walls. It was here, under the watchful gaze of ancient Martian statues, that the Emperor often summoned his sons the Eleven Brothers of Mars.

Each son ruled a district of the Empire, their territories spread across the red planet like pieces on a chessboard. Yet, despite their shared blood, they were far from united. Toxic rivalries, jealousy, and an insatiable hunger for power poisoned their relationships, ensuring that the bonds between them were as fragile as Martian glass.

At the head of the council table sat Emperor Ramses Solis II, his piercing amber eyes scanning the room. His presence was magnetic, his authority absolute. To his left sat Princess Ankhes Solis, her expression unreadable, though her sharp gaze betrayed her disdain for the petty squabbles of her brothers. She sat in silence, a spectator to the drama unfolding around her.

The Eleven Princes

The Emperor's sons were as different as the districts they ruled, yet each carried the unmistakable mark of their father. They were tall, imposing figures, their features chiseled and their eyes the same amber hue that symbolized the Solis bloodline. But beneath their regal exteriors lay a cauldron of ambition, envy, and resentment.

1. Prince Setek Solis (District 1): The eldest and the most ruthless. Setek was the Emperor's right hand, a military genius who commanded the Empire's army with an iron fist. His loyalty to the Emperor was unquestionable, but his disdain for his younger siblings, especially Ankhes, was barely concealed.

2. Prince Horon Solis (District 2): The schemer. Horon was a master of intrigue, ruling the second district with cunning and manipulation. He thrived on chaos, often playing his brothers against one another for his own gain.

3. Prince Ammon Solis (District 3): The scholar. Ammon was the Empire's foremost historian and philosopher, a man more interested in preserving the past than shaping the future. He often clashed with Setek, whose militaristic approach he viewed as barbaric.

4. Prince Dagon Solis (District 4): The zealot. Dagon was deeply religious, seeing himself as a divine instrument of Martian destiny. His fervor made him a polarizing figure, respected by some and feared by others.

5. Prince Khepri Solis (District 5): The innovator. Khepri was a scientist and engineer, responsible for many of the advancements that had shaped the Empire. His inventions were both

a boon and a threat, as he often used them to outmaneuver his siblings.

6. Prince Meren Solis (District 6): The shadow. Meren was a master of espionage, the head of the Empire's intelligence network. He thrived in secrecy, his motives known only to himself and, occasionally, the Emperor.

7. Prince Osiron Solis (District 7): The diplomat. Osiron was the Empire's ambassador, skilled in the art of negotiation. He often acted as a mediator between his warring brothers, though his neutrality earned him little respect.

8. Prince Thutmose Solis (District 8): The gladiator. Thutmose was a warrior at heart, ruling his district with brute force. He was fiercely loyal to Setek, whom he idolized, but his temper often made him a liability.

9. Prince Rahan Solis (District 9): The artist. Rahan was the Empire's most unconventional son, a painter and poet who viewed his role as ruler with disdain. His eccentricity made him a target of ridicule, but his bond with Ankhes was unshakable.

10. Prince Nefer Solis (District 10): The merchant. Nefer controlled the Empire's trade routes and economy, a shrewd businessman whose wealth made him one of the most powerful brothers. His ambitions often clashed with those of Setek and Horon.

11. Prince Kael Solis (The Banished, Districtless): The outcast. Kael's name was spoken in whispers, his legacy a stain on the family's reputation. Though banished to the South, his shadow loomed over the Empire, a constant reminder of the Emperor's failure to control his own blood.

The Council Meeting

The meeting began as it always did, with Setek reporting on the Empire's military strength. His voice was steady and commanding, his words carefully chosen to remind his siblings of his dominance.

"Our forces are unmatched," Setek declared, his amber eyes scanning the room. "The rebels in the South are nothing more than a nuisance. A single strike would end them."

"You underestimate Kael," Ammon interjected, his tone measured. "He knows our weaknesses better than anyone. Dismissing him as a nuisance is a mistake."

Setek glared at his brother. "Your philosophy won't win wars, Ammon. Kael is a traitor, and traitors deserve no mercy."

Ankhes shifted in her seat, her gaze fixed on Setek. "Kael may be a traitor," she said, her voice calm but firm, "but he's also a Solis. We should tread carefully."

Setek's expression darkened. "Careful is a word for cowards."

Before Ankhes could respond, Horon leaned forward, a sly smile on his lips. "Perhaps we should ask the Emperor what he thinks," he said, his tone dripping with mockery.

The Emperor, who had been silent until now, raised a hand. The room fell silent instantly.

"Kael is no longer my concern," Ramses said, his voice cold and detached. "He is a ghost, a relic of the past. Our focus should be on the future of the Empire."

Ankhes' Perspective

Ankhes listened to her father's words, her expression unreadable. But inside, she felt a pang of sadness. Kael was more than a ghost to her. He was her brother, her blood. And despite everything, she couldn't shake the feeling that he was right—that the Empire was broken, rotting from within.

She glanced at Rahan, the only brother she truly trusted. His eyes met hers, and she saw the same doubt reflected in his gaze. They were both outsiders in their own way, bound by their shared disdain for the Empire's corruption.

As the meeting continued, Ankhes found her mind drifting. She thought of Kael's message, the one delivered in secret just days earlier. He had asked her to meet him, to see for herself what he had built in the South. The temptation was overwhelming, but the risk was enormous. If her father or Setek discovered her plans, the consequences would be dire.

Rivalries and Alliances

The meeting ended with the brothers leaving in their usual fashion—bickering, plotting, and vying for their father's favor. Setek and Thutmose left together, their bond as warriors unshakable. Horon lingered, his sharp eyes observing everyone before slipping away. Ammon and Osiron exchanged quiet words, their alliance one of mutual respect. Nefer disappeared without a word, his mind likely on his ever-growing wealth.

Ankhes and Rahan stayed behind, sharing a quiet moment in the empty hall.

"Do you trust any of them?" Rahan asked, his voice soft.

Ankhes shook her head. "No. Do you?"

Rahan smiled faintly. "Only you."

The Emperor's Wrath

As the brothers dispersed, Ramses retreated to his private chambers. Alone, he allowed himself a moment of vulnerability. Kael's name echoed in his mind, a reminder of his greatest failure. He had spared Kael's life out of weakness, and now that weakness threatened to destroy everything he had built.

But Ramses was not a man who dwelled on regret. He would crush Kael, just as he had crushed every other threat to his rule. And if Ankhes or any of his sons stood in his way, they would share the same fate.

Blood of Mars - Chapter 5: Whispers of Rebellion

The red Martian sky hung heavy above the Citadel as Ankhes sat alone in her private chambers, the letter from Kael clutched tightly in her hands. The words scrawled on the parchment burned into her mind. Kael had warned of rebellion, of a weapon so devastating it could bring an end not only to the Empire but to Mars itself.

Her chest felt tight as she read the letter again, her brother's plea ringing clear in her thoughts.

"Ankhes,

The tools our father used to create his empire are the very tools that will destroy it."

Kael's words felt like a knife, cutting into her loyalty to her father. For the first time, Ankhes doubted not only the Emperor's decisions but the future of Mars itself.

A Private Meeting

Ankhes rose from her seat and summoned her new chief of the Princess's Guard, Commander Sefar, a man she trusted implicitly despite the weight of her own secrets. When he arrived, she wasted no time.

"Bring me the lead terraformers and scientists," she ordered. "I want them here within the hour."

Sefar hesitated for a fraction of a second, sensing the urgency in her tone. "Yes, Your Highness," he said, bowing before departing.

Within an hour, the brightest minds of the Empire were gathered in Ankhes' private war room. Holographic displays of terraforming schematics and energy grids filled the space. The engineers and scientists stood nervously, their eyes darting between each other as they awaited the Princess's questions.

"You all know why you're here," Ankhes began, her voice steady but laced with tension. "The Emperor has built this empire on the promise of a fully terraformed Mars. But I need to know the truth about the risks—the dangers we face. Spare me your loyalty to my father. I need honesty."

The lead engineer, Dr. Maren Selik, stepped forward. A gaunt woman with silver streaks in her hair, she exuded the authority of someone who had worked on Mars' terraforming since its inception.

"Your Highness," Maren began, her voice hesitant, "terraforming Mars is inherently unstable. The Genesis Reactors we use to seed the land with life are, at their core, controlled detonations.

The chain reaction they create is precisely calibrated to ignite the atmosphere without causing catastrophic damage."

"And if that calibration is disrupted?" Ankhes pressed.

Maren hesitated, glancing at her colleagues. "If the calibration fails or is intentionally altered, the device could create a cascading reaction. The impact would be… devastating."

"How devastating?" Ankhes demanded.

Maren took a deep breath. "The explosion would release enough energy to destabilize the planet's core. The Martian surface would fracture, triggering massive volcanic eruptions and atmospheric collapse. Mars would become uninhabitable within weeks."

A heavy silence fell over the room. Ankhes felt a chill run down her spine. "And what if someone—say, the rebels—were to weaponize this technology?"

Maren's face grew pale. "It's possible," she admitted. "If they've managed to modify the Genesis Reactor, they could turn it into a weapon of mass destruction. One device would be enough to wipe out an entire district. Multiple devices… they could destroy the entire planet."

The Threat of the Rebels

Ankhes clenched her fists, anger boiling beneath her calm exterior. "Has this possibility been shared with my father?"

"We've informed the Emperor of the risks," Maren said cautiously. "But he dismissed our concerns. He believes the rebels lack the resources and knowledge to replicate our technology."

"Kael knows," Ankhes said quietly, almost to herself. She turned to the holographic map of Mars, her eyes scanning the territories controlled by the rebels. "Kael understands this technology better than anyone. If anyone could turn it into a weapon, it's him."

Her words hung in the air like a curse. The scientists exchanged uneasy glances, none of them willing to contradict the Princess.

An Impossible Dilemma

Ankhes dismissed the scientists, keeping only Commander Sefar in the room. As the door closed, she sank into a chair, her head in her hands.

"What are you thinking, Your Highness?" Sefar asked, his voice soft.

"I'm thinking," Ankhes said, her voice trembling with frustration, "that my father is a fool. Kael warned me, Sefar. He warned me that this rebellion would tear Mars apart. And I didn't listen."

Sefar hesitated. “Your father is stubborn, yes. But he is also the Emperor. You cannot act against him.”

Ankhes looked up at him, her amber eyes blazing. “Can’t I? Do you really think I’m going to sit by and watch Mars burn because of his pride?”

Sefar didn’t respond. He had served the royal family long enough to know when words were futile.

Ankhes stood and paced the room, her mind racing. “We need to stop this,” she said finally. “But if my father finds out I’ve even entertained the idea of negotiating with Kael, he’ll see it as treason.”

“Then we act in secret,” Sefar said. “What do you need me to do?”

Ankhes paused, her thoughts aligning into a plan. “Kael’s message mentioned a meeting. If he’s willing to talk, then so am I. Arrange for my departure to the South. Quietly.”

Sefar’s expression darkened. “You’re walking into enemy territory.”

“I know,” Ankhes said. “But this is bigger than my father. If there’s even a chance to stop this war, I have to take it.”

Sefar nodded reluctantly. “It will be done.”

A World on the Brink

As Ankhes stared at the holographic map, the weight of the rebellion’s threat pressed down on her. The idea of Kael turning the Genesis Reactor into a weapon wasn’t just terrifying—it was catastrophic. If her father refused to see reason, then she would have to act alone.

For the first time in her life, Ankhes felt truly powerless. And yet, within that powerlessness, a spark of determination ignited. She would not let Mars fall. She would find a way to stop this, even if it meant betraying her father.

As she prepared for the journey to the South, Ankhes couldn’t shake the feeling that this was only the beginning. The whispers of rebellion had become a roar, and the Blood of Mars would soon be spilled.

Blood of Mars - Chapter 6: The Southern Emperor

Kael Solis stood atop a frozen ridge, overlooking the sprawling city he had built in the desolate South Pole of Mars. The city's glowing lights pierced the endless night, casting hues of blue and crimson across the icy landscape. Beneath his feet lay the culmination of years of toil, sacrifice, and rebellion: the Southern Empire, a haven for the outcasts and the disenfranchised who had been abandoned by Ramses' tyranny.

In the biting cold, Kael's presence burned with purpose. He was no longer just a banished prince. He was something far greater. And tonight, he would make it known.

A Memory of Wisdom

Kael's thoughts wandered to a memory that had shaped him. Years ago, during one of the rare moments of peace in the royal court, his brother Ammon, the scholar of the family, had spoken to him of philosophy. Kael had been young, brash, and full of ambition, yet Ammon's words had stayed with him.

"Do you know what makes a man truly exist?" Ammon had asked, pacing the library of the Citadel.

Kael had shrugged, uninterested at the time. "Power. Strength."

Ammon had smiled faintly, shaking his head. "Cogito, ergo sum," he had said. "I think, therefore I am."

Now, standing in the icy winds of the South, those words resonated with Kael in a way they never had before. He thought. He existed. He ruled.

And tonight, the entire planet would know it.

The Southern Coronation

In the heart of the Southern city, a grand hall had been constructed—a stark contrast to the cold and barren landscape outside. It was a symbol of defiance, its walls lined with glowing Martian crystals and banners bearing the sigil of the Southern Empire: a crimson phoenix rising from the ashes.

Kael entered the hall to the thunderous applause of his followers. The room was packed with people—Earthlings, hybrids, and Martian defectors alike. They had all come to witness history.

At the far end of the hall stood a simple throne, crafted from obsidian and ice. Kael approached it with measured steps, his cape billowing behind him. As he ascended the dais, the crowd fell silent.

One of his closest allies, General Taron Vex, stepped forward, holding a gleaming Martian blade. "Kneel," Taron said, his voice echoing through the hall.

Kael knelt, bowing his head as Taron held the blade aloft. "By the will of the people, and by the blood of Mars, I name you Kael Solis, Emperor of the South."

The crowd erupted into cheers as Kael rose, the blade now in his hands. He raised it high, his amber eyes scanning the room.

"Today," Kael began, his voice commanding and resolute, "we are no longer exiles. No longer rebels. Today, we are a nation—a people united under one banner. The Southern Empire is born, and with it, a new era for Mars."

The hall shook with applause and chants of "Emperor Kael!" But Kael raised a hand, silencing the crowd.

"Our fight is far from over," he said. "Ramses Solis and his empire of oppression will not yield without a war. But let him know this: Mars has room for only one Emperor. And soon, my father will be no more."

The Message

As the crowd dispersed, Kael retreated to his private chambers. The coronation had been a success, but the next step was crucial. He needed to send a message—not just to his father, but to his brothers and sister. They needed to understand the gravity of what had happened, and the inevitability of what was to come.

Kael summoned one of his spies, Selene, the same woman who had delivered his earlier message to Ankhes. She was his most trusted operative, a shadow within the Empire.

"Are the messages prepared?" Kael asked.

Selene nodded, holding up a small device. "Every brother, the Princess, and the Emperor himself will receive it. It's encoded. They'll know it's from you."

Kael smiled grimly. "Good. Deliver them immediately."

Selene left without another word, disappearing into the night like a wisp of smoke. Kael turned to the map of Mars displayed on the wall, his gaze fixed on the northern territories.

The Message Delivered

The messages arrived simultaneously across the Empire, delivered directly to the recipients' personal communication devices. Each contained the same chilling declaration:

"To the Emperor,

To the Princes, To my beloved sister Ankhes,

There are now two Emperors of Mars. One in the North, and one in the South.

I have united the rebels, the outcasts, the hybrids, and the Earthlings under one banner: the Southern Empire. We will no longer bow to tyranny, nor will we be ignored.

Mars has room for only one Emperor. And soon, Father, you will be dead.

Signed,

Kael Solis, Emperor of the South"

The Aftermath

The message rippled through the Empire like a shockwave. In the Citadel, Ramses Solis read the declaration with a fury that could have ignited the skies. His amber eyes burned as he crushed the device in his hand, the words "Emperor of the South" echoing in his mind.

"Kael dares to call himself Emperor?" Ramses growled. "He is no Emperor. He is a traitor, and he will die a traitor's death."

Setek Solis, the eldest prince, was the first to respond to the message. "This is an act of war," he said, his voice filled with contempt. "Give me the order, Father, and I will crush him."

But Ramses remained silent, his mind racing. Kael's knowledge of the Empire's technology made him more dangerous than any army. The Genesis Reactors, the trade routes, the supply chains—Kael knew them all. And now, he had united the rebels into a force that could challenge the Empire itself.

In her chambers, Ankhes read the message with a mix of fear and hope. Kael's boldness was reckless, but it also inspired her. For the first time, she saw a glimmer of a future where Mars might be united—not under her father's tyranny, but under Kael's vision.

The other brothers reacted predictably, each viewing Kael's declaration as either a threat or an opportunity. Horon saw it as a chance to manipulate the family dynamics to his advantage. Ammon viewed it as a philosophical dilemma, while Nefer worried about its impact on trade. Thutmose, ever the warrior, sharpened his blade, eager for the battle to come.

Kael's Resolve

Back in the South, Kael stood on the balcony of his fortress, staring out at the icy expanse. The declaration had been sent, the line drawn. There was no turning back now.

"Let them come," he muttered to himself, his voice low and resolute. "Let them come and see what the South is capable of."

For the first time in years, Kael felt truly alive. He was no longer the banished prince, no longer the outcast. He was an Emperor. And he would stop at nothing to claim what was rightfully his.

Blood of Mars - Chapter 7: The Emperor's Fury

The Great Hall of the Citadel was shrouded in tension as Emperor Ramses Solis II paced before the grand council table. The air was heavy, charged with the weight of Kael's declaration. The Emperor's amber eyes glowed with a fury that seemed to pulse in the dimly lit chamber. He was seething, his composure unraveling as the weight of his banished son's rebellion threatened to crack his iron rule.

Kael's message had landed like a thunderbolt, igniting the fires of war. But it wasn't just Kael's words that fanned the flames—it was his audacity. To call himself an Emperor, to unite the South under his banner, to openly threaten his father's life—it was an affront Ramses could not tolerate.

"Kael has crossed a line," Ramses growled, his voice a low rumble that echoed off the obsidian walls. "He has declared himself an Emperor. He has united the filth of Mars against us. And he dares to threaten my life."

His sons, seated at the table, exchanged uneasy glances. Setek, the eldest and most loyal, leaned forward, his jaw tight. "Father, give me the order. I will march to the South and crush him. His rebellion will be a lesson to all who defy the Empire."

Ramses stopped pacing, his gaze locking onto Setek. "This is no ordinary rebellion, Setek. Kael knows our weaknesses. He knows the Empire's secrets. This is not just a fight for dominance—it is a fight for survival."

Ankhes' Dilemma

From her seat at the far end of the table, Princess Ankhes Solis listened in silence. Her father's fury was palpable, his anger radiating like heat from a star. But beneath her calm exterior, Ankhes was anything but composed.

Kael's message had shaken her to her core. His words were not just a declaration of war—they were a plea for unity. A warning. And Ankhes couldn't ignore the truth buried within them. She had spent the last week poring over reports from the South, interrogating engineers, and piecing together the reality of the Genesis Reactors. The rebellion's weaponized use of terraforming technology was no idle threat—it was a catastrophe waiting to happen.

"Father," Ankhes began, her voice steady but cautious, "Kael's rebellion is dangerous, yes. But we must not underestimate the consequences of marching to war."

The Emperor turned to her, his eyes narrowing. "And what would you suggest, daughter? That we sit idly by while Kael builds his empire? That we allow him to challenge my rule?"

"I suggest caution," Ankhes said carefully. "Kael's knowledge of our technology makes him a threat unlike any we've faced before. If we march into the South unprepared, we risk playing into his hands."

Setek scoffed, his impatience evident. "This is not the time for caution, Ankhes. This is the time for action."

"And what if that action destroys Mars?" Ankhes snapped, her composure slipping. "What if Kael uses the Genesis Reactors to retaliate? Do you understand what that would mean?"

The room fell silent. Even Setek had no response. The thought of Mars itself being torn apart by the very technology that had built the Empire was too horrifying to dismiss.

The Emperor's Ultimatum

Ramses broke the silence, his voice cold and commanding. "Kael is a traitor. A cancer. And like any cancer, he must be cut out. We will not negotiate with him. We will not show weakness. Mars belongs to the Empire, and I will not let my wayward son take it from me."

Ankhes clenched her fists under the table, her nails digging into her palms. Her father's stubbornness was infuriating, but it was also unshakable. There would be no reasoning with him.

"Ankhes," Ramses continued, his gaze piercing, "you are my heir. It is time you prove your loyalty to the Empire. You will travel to the South as my envoy. You will deliver an ultimatum to Kael: surrender, or be destroyed."

The room erupted into murmurs. Ankhes stared at her father, stunned. "You want me to go to the South? Alone?"

"You will be accompanied by an escort," Ramses said dismissively. "But you will speak for me. Kael must understand that there is no room for negotiation. The Empire's rule is absolute."

Ankhes felt a wave of dread wash over her. This was not just a test of her loyalty—it was a death sentence. If she traveled to the South, she would be walking into enemy territory. And if Kael refused to surrender, she would be caught in the crossfire of a war she was beginning to doubt entirely.

A Choice to Make

As the council meeting adjourned, Ankhes lingered in the hall, her mind racing. Her brothers dispersed, some eager to prepare for war, others retreating to their own schemes and machinations. Only Rahan, the artist-prince, remained behind.

"You're troubled," Rahan said, his voice soft.

Ankhes sighed, her shoulders slumping. "Troubled doesn't begin to describe it."

"You don't have to go," Rahan said. "You could refuse."

"And risk my father's wrath?" Ankhes shook her head. "You know as well as I do that defying him is not an option."

Rahan hesitated. "What if Kael is right? What if the Empire is the real threat?"

Ankhes looked at her brother, surprised by his words. "You sound like a rebel."

"I sound like someone who sees the cracks in the foundation," Rahan said quietly. "Be careful, Ankhes. Whatever choice you make, it will change everything."

The Princess Prepares

That night, Ankhes summoned Commander Sefar to her chambers. As always, her trusted guard stood by her side, his loyalty unwavering.

"Prepare my escort," Ankhes ordered. "We leave for the South at dawn."

Sefar hesitated. "Are you sure this is wise, Your Highness? Kael is unpredictable."

Ankhes met his gaze, her expression resolute. "This isn't about wisdom, Sefar. It's about survival. If my father won't see reason, then I'll have to find another way."

Sefar bowed. "As you wish."

Kael's Response

In the South, Kael received word of Ankhes' impending visit through his network of spies. The news brought a smile to his lips.

"So, she's coming," Kael said, leaning back in his chair. "Good. Let her come. Let her see what the Southern Empire truly is."

General Taron Vex frowned. "Do you trust her?"

“I trust that she’s as trapped by our father as I once was,” Kael said. “But trust? No. Trust is a luxury I can’t afford.”

Kael stood, his gaze fixed on the horizon. “Prepare the city for her arrival. Let her see the South in all its glory. And let her decide where her loyalty truly lies.”

Blood of Mars - Chapter 8: Secrets of the Red Blood

The crimson hues of the Martian sunset filtered through the towering windows of Ankhes' chambers, bathing the room in a soft, otherworldly glow. She sat in front of her ornate mirror, her reflection revealing a storm of emotions she could no longer suppress. Kael's words echoed in her mind, his message delivered days ago now carrying an even deeper weight.

"Mars has room for only one Emperor. Join me, Ankhes, before it's too late. Together, we can unite the planet. Together, we can rule."

The princess had spent days replaying those words, analyzing every inflection, every hidden meaning. But it wasn't just his declaration of war that haunted her—it was the revelation about Kael's true father. Kael, the exiled prince, the rebel, was not her full brother. His blood was hybrid, a mix of Martian and Earthling. He was a symbol of everything her father despised, yet he had risen to become the leader of the Southern rebellion. And now, he had declared himself Emperor of the South.

Ankhes' thoughts twisted into forbidden territory. The idea of Kael—his defiance, his power, his audacity—consumed her. Her mind wandered to memories of their childhood, his sharp intellect, his roguish charm, his magnetic presence. He had always been different, but now she understood why. And she couldn't ignore the pull she felt toward him, a force growing stronger with every passing moment.

The Burden of the Red Blood

Ankhes' lineage, the so-called "pure blood of Mars," had always been a source of pride for her father. Emperor Ramses Solis II preached the superiority of their kind, a race engineered to thrive on Mars, their red blood a marker of their supremacy. But Ankhes had begun to question everything she had been taught.

The truth about Kael's father unraveled the foundation of her identity. Kael was proof that the Martian bloodline was not as pure as the Emperor claimed. If her father had lied about Kael's heritage, what other lies had he told? What other secrets were hidden within the Red Blood of Mars?

Her gaze shifted to the artifacts adorning her chamber—a golden Martian sigil, a ceremonial dagger, and an ancient text inscribed with the Empire's laws. Symbols of power, symbols of control. Ankhes felt the weight of her father's legacy bearing down on her, a suffocating reminder of the life she was expected to lead.

But for the first time, she saw a path forward that wasn't bound by her father's will. A path that led to Kael.

The Forbidden Desire

Ankhes couldn't deny the feelings that had taken root in her heart. Kael was no longer just her brother—he was her equal, her opposite, her destiny. His defiance of their father, his vision for Mars, his raw power—it all drew her to him in ways she couldn't explain.

She remembered the stories she had heard of Kael in the South: how he had united the hybrids and Earthlings, how he had built a new society from the ashes of exile. He was no longer the banished prince. He was an Emperor in his own right, a man who had carved out a place for himself in a world that had tried to destroy him.

And yet, there was something deeper, something primal, about the way she thought of him now. She couldn't help but wonder what it would feel like to stand beside him, to share his throne, to rule Mars together. The thought of him—his strength, his confidence, his undeniable allure—consumed her.

The Decision

That night, as the twin moons of Mars rose into the sky, Ankhes made her decision. She could no longer remain in the North, bound by her father's tyranny. She had to see Kael, to hear his plans, to understand the prophecy he had spoken of. The pull was too strong, the path too clear.

She began packing her belongings, her hands trembling with a mix of fear and excitement. She chose only the items that mattered most: a necklace given to her by her mother, a journal filled with her private thoughts, and a blade she had carried since she was a child. Each item held a piece of her past, a piece of the person she was leaving behind.

Ankhes summoned her most trusted guardsmen, men who had served her loyally for years. She looked each of them in the eye, her voice steady but tinged with sadness.

"This is goodbye," she said. "I am leaving the Citadel tonight. I do not know when—or if—I will return. You have served me well, and I will forever be grateful for your loyalty. But this is a journey I must take alone."

The guardsmen exchanged uneasy glances, but they did not question her. They knew better than to challenge the will of the Princess.

The Secret Departure

As the Citadel slept, Ankhes slipped into the shadows, her cloak pulled tight around her. She had arranged for a private transport to take her South, a ship manned by a trusted pilot who owed her a debt. The journey would be long and dangerous, but Ankhes was resolute.

Before boarding, she turned to Commander Sefar, who had insisted on accompanying her to the transport bay.

"Your Highness," he said, his voice low, "are you sure about this? Once you leave, there is no turning back."

Ankhes met his gaze, her amber eyes filled with determination. "I have never been more sure of anything in my life."

Sefar nodded, his expression conflicted. "Then may the stars guide you."

Ankhes stepped onto the ship, the hum of its engines filling the air. As the transport lifted off, she watched the Citadel grow smaller and smaller, until it was nothing more than a distant speck on the horizon.

A New Path

As the ship hurtled toward the South, Ankhes felt a strange sense of peace wash over her. For the first time in her life, she was free—free from her father's control, free from the expectations of the Empire, free to choose her own destiny.

She thought of Kael, his message, his vision for Mars. She thought of the prophecy he had spoken of, a prophecy she was beginning to believe was real.

Ankhes closed her eyes, a faint smile playing on her lips. She was no longer just the Princess of the North. She was something more. And she was ready to embrace whatever the future held.

Blood of Mars - Chapter 9: Shadows of War

The Southern outpost of Tazari Station lay nestled in the jagged canyons of the Martian **southern highlands**, an isolated but critical stronghold for Kael's forces. It was a hub of operations for the Southern Empire, where supplies and weapons were funneled to nearby rebel settlements. Tonight, its guards were alert but unsuspecting, patrolling under the blood-red glow of Mars' twin moons.

They never saw the North coming.

The Storm Descends

At the head of the Northern battalion was Thutmose Solis. Known as the "Gladiator," Thutmose was a towering figure encased in shimmering, state-of-the-art Martian armor. His presence alone was a weapon of fear, his reputation as a relentless and merciless warrior preceding him.

The assault began without warning. In the darkness, the Northern forces unleashed their most devastating weapon: the Helios Cannons. These railgun-like devices, mounted on sleek, spider-like tanks, fired concentrated beams of molten plasma. The cannons emitted no sound until the beams hit their targets, creating a deafening explosion of vaporized metal and rock. The first blast annihilated Tazari Station's outer defenses, sending plumes of dust and fire into the Martian night.

Southern guards scrambled to respond, shouting orders and activating the station's defenses. Automated turrets whirred to life, and the rebels armed themselves with Gauss Rifles, weapons that fired magnetic slugs capable of penetrating armor. But it was chaos—organized chaos for the North, and sheer panic for the South.

Thutmose marched at the forefront of his army, his ablative armor shields shimmering with a faint blue glow. These shields, created by the Empire's finest engineers, absorbed kinetic energy from bullets and redirected it, making him nearly invincible. Behind him came hundreds of Northern soldiers, their armor similarly equipped with advanced shielding technology.

The Southern rebels, by contrast, were a ragtag force. Many wore outdated armor scavenged from abandoned colonies, their shielding flickering and unreliable. They were outmatched, outgunned, and outnumbered—but they fought with desperation, knowing there would be no mercy if they failed.

Weapons of War

The battlefield was a cacophony of sounds: the hiss of plasma beams, the sharp crack of railguns, the shouts of soldiers, and the screams of the dying. The Northern soldiers wielded Phoenix Blasters, sleek weapons that fired superheated plasma rounds. Each shot glowed like a miniature sun, cutting through the thin Martian atmosphere with a high-pitched whine. The Southern rebels retaliated with Arc Bolts, crude but powerful weapons that discharged electrical blasts capable of frying a soldier's shield generator.

The North's tanks, spider-like machines called Ares Walkers, were terrifying in their precision. They moved with an eerie fluidity, their plasma cannons swiveling to target anything that moved. The Southern rebels had no comparable vehicles—only Sand Skimmers, small, hovercraft-like vehicles designed for speed rather than combat. The skimmers darted between the canyon walls, firing volleys of rockets at the Northern tanks, but the Ares Walkers' shields absorbed the hits with ease.

Above the battlefield, Northern Stratos Drones filled the sky. These autonomous machines rained down precision-guided bombs, their sleek, bird-like designs making them nearly invisible in the dim Martian light. The rebels had no air defenses, and the drones decimated their ranks with ruthless efficiency.

The Butcher of Tazari

Thutmose was a whirlwind of destruction. His weapon of choice was the Solar Blade, a plasma-edged sword that burned with a golden light. The blade could cut through metal as easily as flesh, and Thutmose wielded it with terrifying skill. He moved through the battlefield like a force of nature, his armor shields deflecting bullets as he hacked down anyone in his path.

The Southern rebels tried to regroup, but their commanders were picked off by the Northern snipers, who used Specter Rifles capable of firing from miles away. Leaderless and outmatched, the rebels fell into disarray.

One young rebel, no older than seventeen, tried to charge Thutmose with a makeshift spear. Thutmose caught the weapon with his free hand, snapping it in two before driving his Solar Blade through the boy's chest. He didn't even pause to watch him fall.

"Tazari will burn," Thutmose growled, his voice amplified by his helmet's comm system. "No one escapes."

The Bloodbath

The Northern soldiers showed no mercy. They stormed the outpost with brutal efficiency, cutting down anyone who resisted. Those who tried to surrender were executed on the spot. The ground was littered with the bodies of the dead, their blood mingling with the red dust of Mars.

One rebel, a middle-aged man armed with a Gauss Rifle, managed to take down two Northern soldiers before he was overwhelmed. Thutmose himself delivered the killing blow, slicing through the man's chest with his Solar Blade. As the rebel fell, his eyes met Thutmose's, defiant even in death.

The rebels' final line of defense was a group of engineers who had barricaded themselves inside the station's central command center. They were the ones responsible for maintaining the Genesis Devices, the technology that Kael had turned into a weapon of rebellion. Thutmose knew they couldn't be allowed to live.

"Breach the doors," he ordered.

Northern soldiers placed charges on the barricade, the explosion sending shards of metal and stone flying. The engineers fought valiantly, using whatever tools and weapons they could find, but they were no match for the Northern soldiers. Within minutes, they were all dead.

Thutmose surveyed the carnage, his blade dripping with blood. "Burn it all," he commanded.

The Northern soldiers set fire to the outpost, their flamethrowers spewing streams of liquid fire that consumed everything in their path. The flames rose high into the Martian night, casting an eerie glow over the battlefield.

A Message of Fear

As the last of the flames died down, Thutmose turned to his comm officer. "Send a message to the South," he said. "Let them know what happens to those who defy the Empire."

The message was simple: a video feed of the burning outpost, with Thutmose standing at the forefront, his Solar Blade held high.

"Kael Solis," Thutmose said, his voice cold and menacing. "This is only the beginning. The North is coming for you. And we will not stop until you are dead."

The Aftermath

By dawn, Tazari Station was nothing more than ashes. The Northern forces withdrew, leaving behind a scene of utter devastation. Hundreds of rebels lay dead, their bodies unburied, their weapons scattered across the blood-soaked ground.

In the South, Kael's generals received the message with grim faces. The attack had been a massacre, a brutal reminder of the North's power. And though Kael himself remained calm, his resolve only deepened.

"The North has drawn first blood," he said to his council. "Now, we will show them what the South is capable of."

Blood of Mars - Chapter 10: The Princess' Choice

The Southern Citadel, nestled deep within the ice-slicked canyons of the South Pole, was unlike anything Ankhes had ever seen. Unlike the gleaming opulence of the Northern Empire's Citadel, this fortress was carved into the Martian rock, its exterior hidden beneath layers of snow and ice. It was a symbol of resilience, of survival in the face of the Empire's might. But as Ankhes stepped out of her transport under the cover of darkness, her breath visible in the frozen air, she couldn't help but feel the weight of danger pressing against her chest.

The Southern guards, clad in patched but functional armor, surrounded her immediately. Their weapons were cruder than those of the North, but their resolve was unmistakable. Ankhes held up her hands in surrender, her every movement scrutinized by wary eyes. She had come here willingly, but she knew that didn't mean she would be safe.

"Search her," barked one of the guards, a grizzled man with a scar running down his left cheek.

Ankhes didn't resist as they patted her down, removing the ceremonial dagger she had brought with her and confiscating the small communicator she carried. She was keenly aware of the tension in the air, the murmurs of disbelief among the soldiers.

"She shouldn't be here," one of them whispered.

"It's a trap," another muttered. "It has to be."

A Dangerous Welcome

The council chamber of the Southern Citadel was dimly lit, its walls lined with makeshift banners bearing the sigil of Kael's rebellion: a crimson phoenix rising from the ashes of Mars. Ankhes was ushered into the center of the room, her hands bound in front of her. The council members, a mix of hybrid Martians and Earthlings, stared at her with a mix of fear and suspicion.

"Princess Ankhes Solis," one of them said, his voice trembling slightly. "You have no place here. Your presence is a threat to everything we have built."

"I came to speak with Kael," Ankhes said calmly, meeting their gaze with unwavering confidence. "I am not your enemy."

"You're the Emperor's heir," another council member spat. "Your blood is the blood of oppression. How can we trust you?"

Ankhes straightened, her amber eyes blazing. "You don't have to trust me. But Kael will. And when he does, you'll understand why I'm here."

The room fell silent. The council exchanged uneasy glances before one of them spoke up. "Take her to Kael's chambers. Keep her under guard until he returns."

Imprisoned

Ankhes was escorted through the labyrinthine halls of the Citadel, her every step shadowed by armed guards. When they reached Kael's chambers, the door slid open with a low hiss, revealing a stark room furnished with the bare essentials: a bed, a desk, and a single viewport overlooking the frozen wasteland beyond.

Before she could protest, the guards shoved her inside. "You'll stay here until Kael decides what to do with you," one of them said gruffly.

Ankhes turned to face them, her expression defiant. "And my pilot?"

"Dead," the guard said bluntly. "Your ship's been scuttled. You're not going anywhere."

The door slid shut behind her, and she heard the faint hum of the chamber's shield activating. She was alone, but not truly—there was one guard stationed inside the room, his gaze fixed on her, while three more stood outside.

Ankhes sat down on the edge of the bed, her mind racing. She had known the risks of coming here, but she hadn't anticipated this. The council's distrust, the cold hostility of the guards, the sudden realization that she was entirely at their mercy—it all weighed heavily on her.

The Message of the North

The hours dragged on, and Ankhes found herself unable to sleep. The constant loop of the Northern raid, displayed on the chamber's comm panel, kept her awake. The footage was brutal, unrelenting. Flames consumed the Southern outpost, bodies littered the ground, and Thutmose's voice echoed through the room:

"Kael Solis, this is only the beginning. The North is coming for you. And we will not stop until you are dead."

The message played over and over, each repetition twisting Ankhes' emotions into knots. She hated her brother Thutmose for his cruelty, but she also understood the grim efficiency of his actions. The raid had been a warning, a show of power designed to break the rebellion's spirit.

And yet, Ankhes couldn't shake the feeling that Kael had expected this. That he had wanted this.

The Sleepless Hours

As the night stretched on, the exhaustion began to take its toll. Ankhes sat on the floor, her back against the wall, her eyes heavy but unable to close. The guard inside the room watched her silently, his expression unreadable. The hum of the shield and the flicker of the comm panel were the only sounds, creating an oppressive sense of isolation.

At dawn, the comm panel finally went dark, the silence almost deafening in its wake. Ankhes took a deep breath, her body aching from hours of tension. She closed her eyes for a brief moment, savoring the stillness.

Then, the door to the chamber slid open.

Kael's Arrival

Kael Solis stepped inside, his presence filling the room like a storm. He was taller than she remembered, his dark hair disheveled, his face marked with the grime of battle. His eyes, sharp and piercing, locked onto hers with an intensity that made her heart skip a beat.

For a moment, neither of them spoke. The weight of their shared history, their complicated bloodline, and the rebellion that now defined their lives hung heavy in the air.

"Ankhes," Kael said finally, his voice low but steady. "You shouldn't have come."

She stood, her legs unsteady but her resolve firm. "And yet, here I am."

Kael studied her for a long moment, his expression unreadable. Then, without another word, he dismissed the guard with a wave of his hand. The shield deactivated, and the door slid shut behind him.

They were alone.

To Be Continued...

Throne of Ash - Chapter 1: The Emperor's Demand

The Southern Throne Room

The air inside Kael's chamber was thick with tension, the weight of unspoken words pressing down on Ankhes like the Martian gravity itself. The flickering glow of the bio-fueled sconces cast jagged shadows along the walls, making the already imposing space feel even more oppressive. This was not a place of comfort. It was a battlefield of the mind, a throne room for a ruler forged in exile, and now—the home of her greatest enemy.

She stood before Kael Solis, stripped of her weapons, her transport scuttled, her presence met with suspicion and hostility. The council had already debated her fate—most had voted for execution. A Northern royal could not be trusted, and yet Kael had overridden them. He had other plans.

Kael leaned back in his chair, dressed in the layered combat robes of the South, his dark hair damp from the frost that still clung to the fabric. He had come straight from the battlefield, his face marked by war, his eyes gleaming with a mix of calculation and hunger—but for what, Ankhes was not certain.

Then came his first command.

"Strip."

Submission or Death

Ankhes' breath hitched, her heart hammering in her chest. She had expected interrogation, perhaps even torture—but not this.

She held her chin high. "What?"

Kael tilted his head slightly, studying her, testing her resolve like a predator circling prey. "I need to know who you really are. If you carry any marks of the Empire—tattoos, insignias, trackers—I must see them." His voice was low, dangerous, but not without amusement.

Her mind raced. Was this merely a security measure? A game? A display of power? Perhaps all three.

Kael rose from his throne, taking slow, measured steps toward her. "You have two choices, Ankhes. You can die here, in this room, alone and forgotten. Or…" His voice dipped lower, a blade pressing against her resolve. "*You kneel, you submit, and you become my Empress.*"

The air between them felt electrified.

Ankhes' body burned with defiance, but her mind whispered this was not defeat—this was an opening. A way into Kael's world, into his war. Her decision had already been made the moment she stepped off her ship. She wasn't here to fight him.

She was here to own him.

Slowly, she reached for the clasps of her uniform, keeping her gaze locked on his. If he wanted a test of loyalty, she would give him one. But on her terms.

The Confession

As the heavy fabric pooled around her feet, Kael's expression remained unreadable. He took a step closer, then another, his fingers tracing a slow path along the bare skin of her collarbone—not with desire, but curiosity. Searching.

"No marks," he murmured.

"You expected a brand of the Empire?" she challenged.

"I expected a liar," he countered.

Ankhes allowed a ghost of a smirk. "Then you expected wrong."

Kael's hand dropped, his gaze locked onto hers. "Then tell me, Ankhes Solis… why are you really here?"

She could have lied. She could have fed him a carefully rehearsed script about diplomacy, about peace. But that was not why she had left her father's citadel in the dead of night, why she had crossed into enemy territory alone.

Because Kael was not her enemy.

"Because you are not my brother."

Kael's breath stilled.

The silence between them was deafening.

She saw it then—the moment his mind began reconstructing reality, the cracks forming in the foundation of the lies he had been told. His fists clenched at his sides, his entire body coiled with restrained fury.

"Explain."

And so she did.

The Emperor's Response

Ankhes did not falter, did not break eye contact as she revealed the truth. How Kael had been raised in the shadow of a lie, how his exile had been orchestrated to ensure the Emperor's throne remained unchallenged. How the blood he carried was not that of the Empire, but of a forgotten truth buried in the sands of Mars.

Kael listened, his jaw tightening, his rage simmering beneath the surface like a dormant supervolcano.

When she finished, he exhaled slowly, controlled.

"Then it is settled," he said at last, his voice stripped of all doubt. "You will not return to the North. You are mine now. Mine to keep, mine to protect, and mine to rule beside me."

Ankhes lifted her chin. "And if I refuse?"

Kael smirked, dark and knowing. "Then you wouldn't have come here in the first place."

And she hadn't.

The Southern War Council

As dawn painted the frozen Martian horizon in streaks of gold and crimson, the Southern Citadel roared to life.

Kael's advisors, generals, and warlords gathered in the grand hall, their voices urgent as reports flooded in. The North had struck again, destroying an outpost near the equatorial belt—one step closer to the heart of the rebellion.

Kael stood before them, his black combat cloak draped over his shoulders, his presence an unshakable force.

He raised a hand, and the room fell silent.

"The time for patience is over," Kael declared. "Our enemies believe they have broken us. They think we will cower and beg for mercy." His voice turned sharp, lethal. "But today, we send them a message."

A low murmur spread through the room, tension thickening.

Kael turned to one of his officers. "Prepare the first launch."

The Terraforming Ballistic Weapon—Project Genesis—was still considered unstable. A bomb that, if deployed correctly, could reshape entire landscapes, igniting chain reactions that turned barren rock into living soil—or, if weaponized, obliterated entire strongholds.

The council hesitated.

Ankhes, standing beside Kael now as his equal, felt the moment before it arrived. The shift in control. The silent acknowledgment that Kael Solis was no longer the exiled prince.

He was the Southern Emperor.

He turned to her then, his voice dropping into something only she could hear.

"Tell me, my Empress," he murmured, challenging her. "Where shall we send our first gift?"

A slow, knowing smile curved on Ankhes' lips.

She had chosen her side.

And the war had just begun.

Throne of Ash - Chapter 2: The Emperor's Fire

The War Council

The chamber was filled with shadow and fire, the flickering glow of the holo-maps casting eerie, shifting light across the gathered warlords. The Council of the South had convened in urgency, every voice sharp, every hand ready to draw blood.

At the head of the room, Kael Solis stood motionless, drinking in the fear and anticipation of those sworn to him. His black robes whispered against the stone as he stepped forward, raising a single hand.

Silence.

“This night,” Kael said, voice edged with steel, “the North learns what it means to wage war against the South.”

The table before them glowed with a holographic map of Mars, the red planet split between dominance and defiance. The Northern Empire, pristine and powerful, glittered like a venomous jewel in the sunlight. The South, its rebellious and raw counterpart, was carved into the canyons, hidden beneath the ice and stone.

Kael tapped the table, and three red markers pulsed to life.

“Three options,” his chief strategist rumbled. “Three districts of the Empire, each vulnerable in different ways. We strike one.”

Option One: The Industrial District (Amun-Seti)

- Weakness: Supplies, Weapons Manufacturing
- Tactical Advantage: Cripples the North’s ability to maintain war machines.
- Risk: Civilian casualties, potential backlash.

Option Two: The Blood Vaults (Etemenanki)

- Weakness: Genetic Engineering, Cloning, and High-Bloodlines
- Tactical Advantage: Destroys the next generation of Northern rulers, weakening their future.
- Risk: Political fallout. Would unite the North in vengeance.

Option Three: The Command Citadel (Ra'Setekh)

- Weakness: Military Leadership
- Tactical Advantage: Striking the heart of command, throwing their forces into chaos.
- Risk: Heavily guarded, high chance of failure without full-scale invasion.

Ankhes' Decision

The council erupted into argument. Some wanted blood. Some wanted control. Some wanted complete annihilation.

Kael said nothing, watching his Empress.

Ankhes, standing beside him now, felt the weight of expectation. She had not yet earned their trust. Not yet proven her loyalty. But this was her moment.

And she knew exactly what Kael was doing.

"You want me to decide."

A slow, dark smile curled at Kael's lips. "You came here of your own will, my Empress. Now prove you belong here."

Her blood burned with the weight of power.

The Blood Vaults. The lifeblood of the Northern Empire. The future of Mars. Destroying them would end her father's dynasty before it could be reborn.

She lifted her chin. "We strike Etemenanki."

The room stilled.

Kael's eyes flashed with something dark and knowing, and he let out a slow breath, nodding.

"Then it is done."

The Genesis Bomb

The Genesis Reactor was a thing of beauty and terror.

Originally designed to terraform Mars, to bring oxygen and life to the wastelands, Kael had turned it into a weapon.

A controlled detonation of Genesis radiation could melt glaciers, turn rock to fertile soil, and carve oceans into the planet's surface.

A full detonation?

It would reduce an entire city to ash and glass.

"Prepare the launch sequence," Kael commanded. "Etemenanki falls before dawn."

The Sound of War

As the warlords moved to carry out their orders, Kael turned to his second-in-command.

"Open the city's speakers," he said.

The commander hesitated. "What message shall we play?"

Kael's smirk was pure malice.

"Let them hear us. Let them feel us. Let them know that their time is over."

A signal was sent across the Southern Empire.

Every speaker, every city tower, every underground district—even the tunnels where rebels still hid in fear—all of them were flooded with sound.

Not words. Not a battle cry.

Music.

Heavy, deep, throbbing. Rhythms that pulsed with raw, primal power.

A war beat.

Across the South, soldiers grinned, rebels lifted their weapons, and the people of the resistance felt the bass shake their bones.

The North would hear it, too.

Let them tremble.

The Emperor's Prize

Ankhes had never felt so alive.

The sheer power of the moment—the war machine she had set in motion, the fire in Kael's eyes as he commanded the fall of her former home, the thrum of the music vibrating through the citadel—it ignited something in her.

Kael turned to her, eyes dark with something dangerous. "You did well, my Empress."

The title sent a thrill down her spine.

But she wasn't done proving herself.

As the council dispersed, Kael barely had time to react before Ankhes grabbed him by the collar and pulled him into a fierce, bruising kiss.

He growled against her lips, catching her waist, slamming her back against the war table.

"You're playing with fire," he murmured, voice thick with heat.

"Then burn me," she whispered.

Kael did not hesitate.

His hands roamed over her, rough, claiming. She gasped as he lifted her onto the table, shoving aside holoscreens and maps, knocking over goblets of Martian wine.

His lips traveled down her throat, biting, teasing, branding her as his.

Outside, the music thundered.

Inside, she became his.

The war had begun.

Throne of Ash - Chapter 3: Fire from the Sky

Kael sat at the war table, his fingertips pressed together as he watched the countdown on the holo-screen. Around him, his council stood in tense silence. The first Ballistic Genesis Warhead had been armed, primed, and prepared for its maiden flight. There was no turning back now.

The Southern Emperor exhaled deeply, then reached for the communicator.

"Patch me through to the North," he commanded. The transmission was sent across encrypted channels, bouncing between satellites before cracking through every comm device in the Northern Empire—military outposts, citadels, private comms of the royal family. Every Solis son. Every general. And the Emperor himself.

A hush fell across the North as Kael's voice filled every room.

"Go outside."

It was a simple command, but it carried the weight of a death sentence.

Soldiers rushed from their barracks. Civilians spilled into the streets. The Northern princes, skeptical but obedient, stepped onto their palace balconies. The Emperor himself, draped in crimson robes, strode onto the grand terrace of the Imperial Citadel, his golden eyes narrowing as he turned to the sky.

And then they heard it.

A whistle.

It started as a low howl, barely audible, carried by the frigid Martian wind. But it grew. It grew until it became a piercing, terrifying shriek, the sound of death itself. The sound of inevitability. It was unmistakable—the wail of a Ballistic Genesis missile descending from the heavens, its whistling scream designed to paralyze its victims with fear before the inevitable.

Kael leaned back in his chair, watching the telemetry as the warhead plummeted toward District 6.

He activated the second broadcast.

"The South has something the North could only dream of… an Empress."

The transmission flickered for a moment before displaying a single image—Kael, seated at the war table, the princess straddling him, her bare back glistening in the dim light of the command center. Her hands gripping his shoulders. His lips grazing her throat. The image lingered for just long enough before cutting back to static.

A calculated move. A declaration of power. A taunt soaked in arrogance.

Then, the world ended.

A blinding light swallowed the horizon. It erupted into a column of fire, searing through the Martian sky like a newborn sun. The ground rippled outward in a shockwave that shattered glass, cracked stone, and turned men to vapor where they stood. District 6 ceased to exist—not bombed, not destroyed, but erased.

The firestorm expanded in all directions, flattening everything in a radius far larger than predicted. The Genesis Bomb wasn't just a weapon of war; it was an act of godly wrath. The energy release of the detonation rivaled the greatest human-made explosions in history—far beyond conventional nuclear arms, its terraformation core unleashing an unholy cocktail of radiation, plasma, and kinetic annihilation.

Where once stood towering domes, bustling streets, and sprawling industrial sectors—now lay an ocean of black glass and charred bone.The air itself shimmered with unnatural heat, warping reality in its wake. The fallout would spread for weeks. One million lives, gone in an instant.

The Northern Citadel stood in stunned silence. For the first time in his reign, Emperor Ramses Solis II did not speak.

Kael let the silence linger before delivering the killing blow.

"Your Excellency…"

A new transmission opened. The princess appeared on-screen, her bare skin now hidden beneath ceremonial robes, her expression void of defiance. She knelt before the comm, her hands clasped.

"Father… please." Her voice trembled, but her eyes burned with something dangerous. "Stop this. End it. There is no Empire without its people. There will be nothing left if this continues. The next Genesis Bomb is armed. Two more will follow. And the fourth…"**

She hesitated, swallowing.

"…Will be aimed at your throne."

The North was paralyzed. The Emperor clenched his fists so tightly his nails dug into his palm. His sons stood behind him, pale, silent, knowing they had just witnessed the impossible.

Kael ended the transmission with one final warning.

"You have one hour to sign the treaty. Or your palace will be a Throne of Ash."

The war had changed. Forever.

Throne of Ash - Chapter 4: The Emperor Falls

The throne room of the Northern Citadel had never felt so quiet. The air was thick with incense, the torches casting long shadows across the obsidian walls. The once-mighty Emperor Ramses Solis II sat slumped in his chair, his golden robes weighed down by the knowledge that his empire was crumbling around him. His once-unshakable rule had been reduced to whispers of treason, rumors of betrayal, and a looming war that he no longer had the strength to fight.

He had seen the blast from Kael's weapon, felt the tremors beneath his feet. The Genesis Bomb had changed everything. A million lives, erased in an instant. The South had made its message clear—Mars was no longer his.

But that wasn't what haunted him most.

It was the silence of his sons.

They had once sat beside him in court, armored in their blood-red Martian cloaks, sworn to protect the empire that bore their name. Now? They were scattered, lurking in the shadows, plotting in secret. None of them had sworn loyalty after the bombing. Not one.

He had summoned them here, to his throne room, to demand their allegiance. He had expected defiance. But he had not expected Thutmose.

The Betrayal

The doors to the chamber creaked open, and Prince Thutmose entered without hesitation. His armor, normally polished to a gleaming crimson, was splattered with dried blood—not from the battlefield, but from the halls of the Citadel itself.

Behind him, a trail of bodies marked his path. The royal guards lay dead, their throats slit before they could even raise their weapons. The Emperor's council—advisors who had sat at his side for decades—lay slumped against the marble walls, their eyes still wide with shock.

There was no declaration of war. No formal challenge.

Only Thutmose, standing before his father, gripping a dagger still dripping with blood.

"You are weak," Thutmose said. His voice was steady, his expression unreadable. "You should have died long before today."

The Emperor struggled to rise, his once-mighty frame now frail, his hands trembling with age. "You dare—"

"I dare," Thutmose interrupted. "You sat in this throne, believing you were a god, but you let the South grow strong. You let Kael humiliate you. You let Earth see us as prey. I will not allow this empire to rot under your rule."

Ramses reached for his sword. He wasn't fast enough.

Thutmose lunged forward, the dagger flashing under the golden torchlight. The blade sank deep into the Emperor's chest, just below the ribs, twisting with precision. A choked gasp escaped the old man's lips as his fingers clawed at his son's arm.

Thutmose leaned in close, whispering, "You were never strong enough to rule Mars."

He pushed the blade deeper.

The Emperor choked on his own blood. His body convulsed, his golden robes staining red. And then, with a final gasp, Ramses Solis II—the greatest ruler Mars had ever known—was dead.

The North Fractures

By dawn, the Northern Empire was unrecognizable.

With Ramses dead, his eleven remaining sons turned on one another like starving beasts. Each one believed themselves the rightful heir. Each one sought the throne.

But there could only be one Emperor.

The Citadel became a warzone. The royal palace, once the seat of Martian power, was now a battlefield, its corridors running red with the blood of princes. Some fought with blades, others with plasma rifles, and a few with nothing but their bare hands.

By the time the sun set, five of the princes were dead.

Thutmose emerged from the carnage the victor. He had slaughtered his brothers, purged the remaining loyalists, and declared himself Emperor of the North.

But the throne he now sat on wasn't stable.

The surviving princes fled, each one retreating to their own districts, declaring themselves kings of their own fractured territories. The once-mighty Northern Empire was no more. In its place stood dozens of warring factions, each one vying for power, each one waiting for the moment to strike.

Mars had become a battleground.

And Earth was watching.

The Arrival of Earth's War Fleet

High above Mars, the first warships arrived.

Earth's fleet was silent, black against the void, their weapons armed and ready. They had watched as Mars descended into civil war, and now they saw an opportunity.

No unified empire meant no resistance.

Mars was ripe for conquest.

From the bridges of their ships, Earth's commanders observed the chaos below. They saw Thutmose claim the throne. They saw the fractured North. They saw Kael in the South, holding his empire together.

And they decided it was time to act.

The first warships descended into Martian orbit. Landing parties were deployed to key regions, not with the intention of peace, but to seize control.

The war for Mars was no longer just Martians vs. Martians.

Now, Earth was in the game.

Kael's Response

News of Ramses' death reached the Southern Citadel within hours. Kael listened to the reports in silence, his fingers tapping against the war table.

"Thutmose took the throne," one of his advisors said. "He killed the Emperor. The North is fractured—each prince has taken a district."

Kael barely reacted. He had seen this coming.

"And Earth?" he asked.

"They've landed troops in three territories. They're claiming 'peacekeeping operations,' but—"

Kael scoffed. "They're here to take the planet."

Silence filled the chamber. The council members exchanged uneasy glances.

Finally, Kael stood. His emerald eyes burned with a quiet intensity.

"Then they've made a mistake," he said.

He turned to Ankhes, his Empress, who had been standing at his side, listening.

"You wanted peace," he told her. "You wanted unity. But Earth doesn't want peace, and the North doesn't want unity. If we don't act now, there will be nothing left of Mars."

Ankhes met his gaze, her expression unreadable. She knew what he was saying.

They would have to crush Earth's invasion before it could begin.

And to do that?

They would have to unite the South against an enemy greater than Thutmose.

Kael turned back to the council. His voice was cold, calculated.

"Prepare our forces," he ordered. "The war for Mars has begun."

The Cracked Earth

Far from the Citadel, in the wastelands of the Genesis Bomb's explosion, something moved.

The once-molten black glass at the center of the blast zone had cooled into jagged obsidian, smooth as a mirror. But now, it shifted.Fractures snaked across the surface, splitting apart with a sound like distant thunder.

Beneath the glass, something ancient stirred.

Something that had been buried beneath Mars for thousands of years.

And now, it was waking up.

The Silent Ones

A terrifying, unknowable force—neither biological nor technological. When the bomb went off, it opened a rift. Now, something is coming through. Shadows move where there should be none. People disappear without a trace. Communications go silent. Those who venture too close to the blast site report hearing whispers in their minds—not in any language they know, but one they somehow understand.

The war for Mars may have just become a war for survival.

The Bloodborne Curse

The radiation from the explosion doesn't just kill—it mutates. Those who were at the edge of the blast don't die… they transform. Their skin blackens, their eyes burn with an unnatural light, and their minds fracture, filled with a hunger they cannot explain. Some whisper that they are not alive, nor dead—just something else.

Now they roam, spreading the mutation like a plague. Every outpost near the blast zone has gone dark. No one knows how far the infection will spread, but one thing is clear—Mars is running out of time.

The Howling Rift

The Martian ground is no longer silent. A low, guttural hum emanates from the blast zone, resonating through the planet's crust. Scientists say it's just shifting tectonic plates. The soldiers stationed nearby aren't so sure.

They hear screams at night—long, distant, and inhuman. The winds no longer move as they once did. The stars above the impact site seem wrong.

Something else is beneath Mars. And the bomb may have just woken it up.

The Red Eclipse

A dark shape now orbits Mars—too small to be a natural moon, too large to be debris. It wasn't there before the bomb detonated. Astronomers on both sides of the war report their sensors glitching when they try to scan it. No signals come from it, yet it moves as if it's watching.

Some say it's Earth's doing. Others believe it's something older.

And then, without warning, Mars experiences its first eclipse in recorded history.

The sky darkens. The cold intensifies. And in the silence, every radio frequency across the planet goes dead.

Final Words of the Chapter

The Emperor was dead.

The North was broken.

Earth had arrived.

And Kael Solis, the banished prince turned ruler, was the only one left standing between Mars and complete annihilation.

To Be Continued...

Made in the USA
Monee, IL
04 November 2026

67385796R00035